Published 2012 by Cliff Top Press Ltd.
www.arithmeticvillage.com

ISBN 978-0-9845731-5-8

At Arithmetic Village we believe in some fundamental principles. One, that information is a human right and two, that if you ask you shall receive. We still assert the moral right of the author and illustrator.

Printed by Lightning Source, USA

Files licensed by www.depositphotos.com: Leather © montego • Background of burlap hessian sacking © odua • Cardboard background © Leonardi • Green floral fabric © Andriuss • Green harvest - seamless pattern © miaou-miaou • Fabric with red and pinks roses © mirabellart • Knit texture © Goodday • Vintage pattern © Chamille White • Art vintage floral seamless pattern background © Irina_QQQ • Seamless yellow fabric texture © ratselmeister

Arithmetic Village

By Kimberly Moore

To Jim

In Arithmetic Village, there are only two rules,
be kind to each other and share all the jewels.

Magical jewels tumble down from the sky

every Tuesday and no one knows why!

The villagers differ in every way,

But they all love jewels
and all like to play.

Polly collects slow,

Tina collects fast,

Linus always loses, and comes home last!

When the jewels become too tricky to carry,

they visit their friend, a tailor named Mary.

She sews sturdy sacks in her wee little den.

one sack for ten gems!

🪪 = 💎💎💎💎💎💎💎💎💎💎

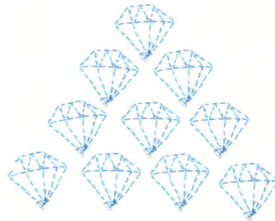

They're crafted with care
so they each can fit ten.

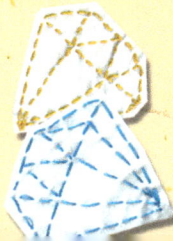

When the sacks become too heavy to hold,

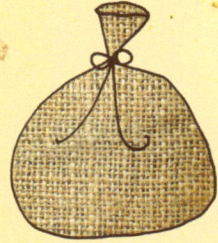

they fit ten in a chest sparkly and gold!

The jewels are gifts for the King in the end.

He's not just the King, he's everyone's friend!

This village is enchanted, the more you return,
the more people you meet,
and the more you can learn!

Dear Grownup,

This book helps children develop a natural understanding of number placement, a basic, yet fundamental component to mastering other mathematical concepts.

To solidify the concept, encourage your child to decorate a personal treasure chest. Collect 100 objects and 10 sacks to place in your treasure chest. You have now created a delightful, personal math manipulative kit!

A light hearted approach to learning, can ignite a lifetime love of math!

For more inspirational ideas visit www.arithmeticvillage.com

Kind wishes,

Kimberly

Arithmetic Village

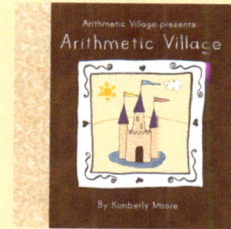

Polly Plus

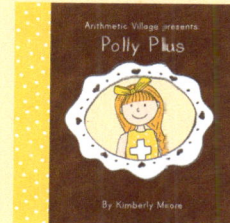

Linus Minus

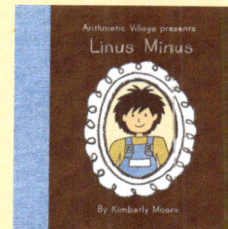

Tina Times

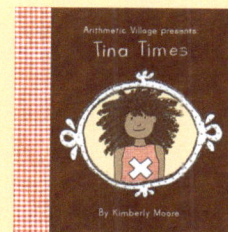

King David Divide

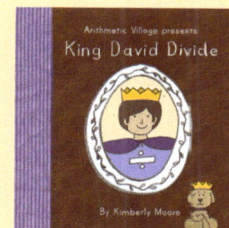

www.ingramcontent.com/pod-product-compliance
Lightning Source LLC
Chambersburg PA
CBHW040025050426
42452CB00003B/146